Birds of Prey

VULTURES

JANET DAVIS-CASTRO

BLACK RABBIT BOOKS

Bolt is published by Black Rabbit Books
P.O. Box 227, Mankato, Minnesota, 56002
www.blackrabbitbooks.com

Marysa Storm, editor; Michael Sellner, designer and photo researcher

Library of Congress Cataloging-in-Publication Data
Names: Davis-Castro, Janet, author.
Title: Vultures / Janet Davis-Castro.
Description: Mankato, Minnesota : Black Rabbit Books, [2023] | Series: Bolt. Birds of prey | Includes bibliographical references and index. | Audience: Ages 8-12 | Audience: Grades 4-6 | Summary: "Equipped with a hooked beak and sharp talons perfect for tearing into carrion, vultures rule the skies. They also capture the imagination of readers everywhere. With an engaging design focused around expertly leveled text and colorful infographics that support visual literacy, introduce reluctant and beginning readers to the life cycle, features, and diets of these powerful birds of prey. Provided by publisher.
Identifiers: LCCN 2021002965 (print) | LCCN 2021002966 (ebook) | ISBN 9781623105570 (hardcover) | ISBN 9781644664889 (paperback) | ISBN 9781623105631 (ebook)
Subjects: LCSH: Vultures–Juvenile literature.
Classification: LCC QL696.F32 D377 2023 (print) | LCC QL696.F32 (ebook) | DDC 598.9/2-dc23
LC record available at https://lccn.loc.gov/2021002965
LC ebook record available at https://lccn.loc.gov/2021002966

Image Credits
123RF: wollertz, 18; Agefotostock: Anup Shah/AnimalsAnimals, 14; Alamy: Albert Bastius, 10–11; dpa picture alliance, 20–21; Malcolm Schuyl, 12; Mark Miller Wildlife Imagery, 28–29; REUTERS, 21; AnimalsAnimals: Ardea/Weaving, Alan, 14; Dreamstime: Dekanaryas, 16–17; Sara Thornton, 4–5; iStock: Kirk Wester, 5; Ullimi, 4–5; Zigmunds Dizgalvis, 6–7; Science Source: Peter Chadwick, 26–27; Shutterstock: Body Stock, 24; digitmilk, 7, 16–17; Dirk M. de Boer, 24; Eric Isselee, 8–9, 23; FotoRequest, 23; Greens and Blues, 19; GUDKOV ANDREY, 13; JIANG HONGYAN, 23; julia badeeva, 6; Lucian Coman, 19; Martin Pelanek, 14; milart, 23; Nick Fox, 32; NRuArg, Cover; PACO COMO, 1, 31; Philippe Clement, 3; Rudmer Zwerver, 24–25; Subhartha Sarkar17, 7; Svetsol, 7, 16–17; volkova natalia, 29

Contents

CHAPTER 1

FLYING High

Turkey vultures circle high in the sky. Something on the ground stinks. The vultures drop down to where the smell is strongest. They land next to a dead animal. Immediately, they begin to tear at the animal's rotting flesh. Time to feast!

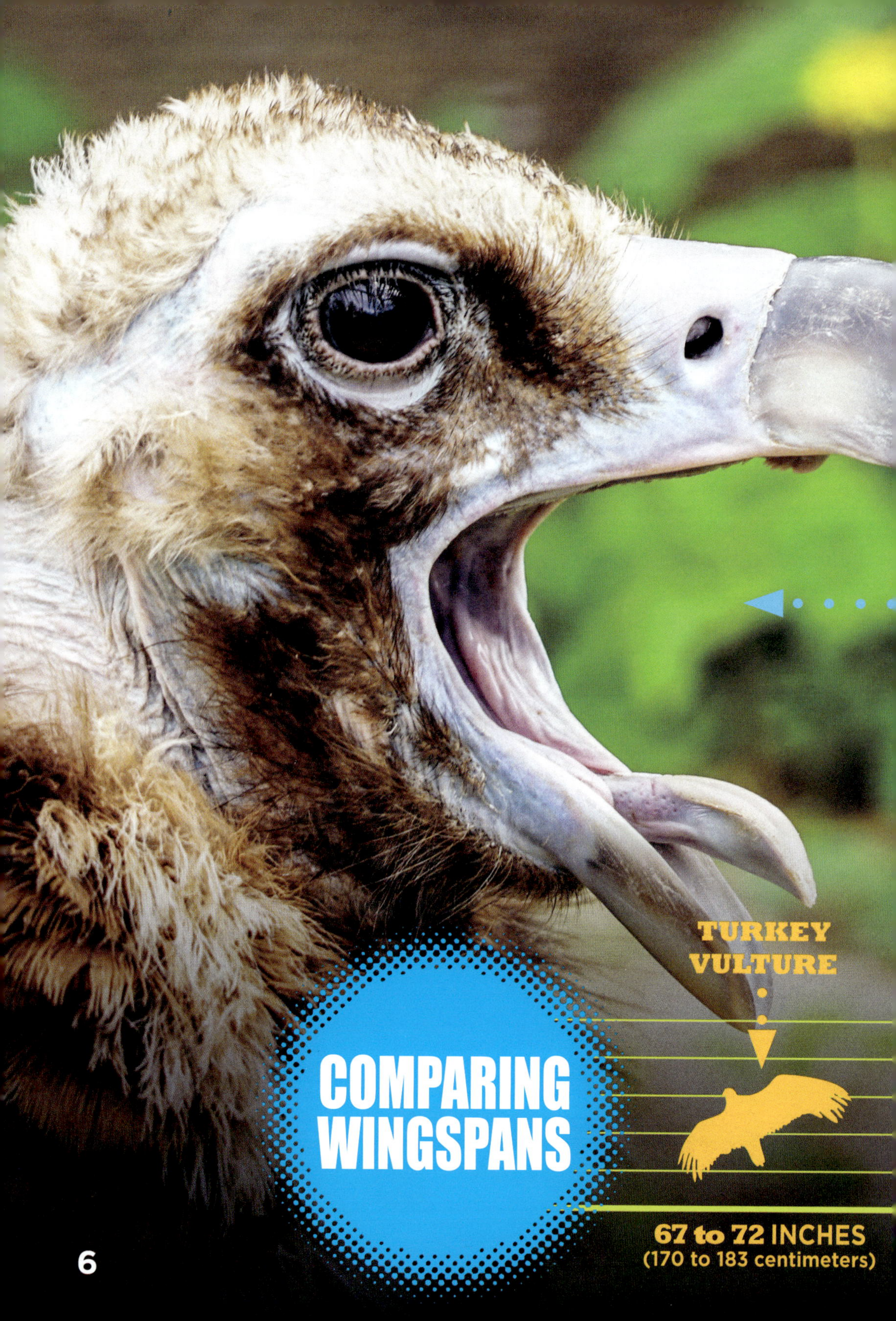

COMPARING WINGSPANS

TURKEY VULTURE

67 to 72 INCHES
(170 to 183 centimeters)

Many Kinds

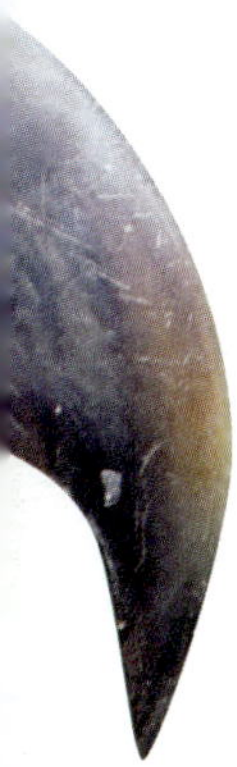

Vultures are birds of **prey**. They eat other animals. There are more than 20 types of these birds. Their featherless heads make them easy to recognize.

There are two main groups of vultures. Old World vultures live in Europe, Asia, and Africa. They squeal and whistle. New World vultures are found in the Americas. They have no voice box. They only grunt and hiss.

PARTS OF A VULTURE

WINGS
TALONS

CHAPTER 2

At HOME and on the Hunt

Unlike other birds of prey, vultures rarely kill their meals. Instead, they're scavengers that eat dead animals they find. Often, these animals died from sickness. Cars hit and killed others. Some vultures steal food from other animals. Occasionally, they'll eat bugs or poop.

Vultures' stomachs have strong acids that kill bacteria. The acids keep them from getting sick after eating rotten meat.

Made to Rip and Tear

These birds' body parts help them eat. Their long necks are great for reaching inside **carcasses**. Their strong beaks help them tear muscle and skin. Most vultures have featherless heads and necks. Having no feathers makes cleanup easier after messy meals.

Habitats

Vultures live all around the world. They make their homes in deserts, forests, grasslands, and mountains. Some even live near cities. These birds spend most of their time in warm spots. Some **migrate** to warmer places in the winter.

Vultures have amazing eyesight. Some can spot food from 2 to 4 miles (3 to 6 kilometers) away.

FORESTS

GRASSLANDS

DESERTS

VULTURE RANGE MAP

This map shows the continents where Old and New World vultures live.

EUROPE
ASIA
AFRICA
OLD WORLD VULTURES

CHAPTER 3

VULTURE Families

Most vultures **roost** in groups. In spring, males and females join together to have babies.

These pairs choose protected spots to lay eggs. New World vultures often pick rocky ledges or holes in trees. Some choose abandoned buildings. Old World vultures build simple nests in similar spots.

Where Vultures Lay Eggs

CLIFFS

TREES

GROUND

Chicks

Once nesting spots are ready, females lay one to three eggs. Parents take turns sitting on the eggs to keep them warm. The babies hatch in about one month.

Chicks are small. They're helpless. They soon get bigger, though. At around three months old, they learn to fly. Most young vultures are ready to find food on their own soon after.

CHAPTER 4

PREDATORS and Other Dangers

Adult vultures are big birds. Their size protects them from most attackers. They have special defenses too. If threatened, the birds throw up. The vomit is stinky and makes many **predators** run away. Throwing up makes these birds lighter too. It's easier to fly away. Predators eat vulture eggs and chicks, though. Other meat eaters steal food from vultures too.

Vulture Food Chain

This food chain shows what eats vulture chicks. It also shows what adult vultures eat.

Some people hunt with lead bullets. Lead is **toxic**. Vultures that eat animals killed by those bullets die.

Human Threats

Vultures face their greatest threats from humans. Farmers will give certain medicine to their cattle. Vultures then die from eating those animals. People also destroy vulture habitats. The birds have less land to live on. Some vultures are now in danger of dying out.

Keeping Them Safe

People often call vultures disgusting. They say the birds are creepy. But vultures have a big job. They clean up dead animals that would spread disease. They are necessary for a healthy planet.

People must protect them.

The first Saturday in September is International Vulture Awareness Day. Zoos and bird centers celebrate by teaching people about vultures.

By the NUMBERS

30 MILES (48 KM) PER HOUR

average flying speed of a California condor

7

NUMBER OF NEW WORLD VULTURE SPECIES

up to **33**
POUNDS
(15 kilograms)
WEIGHT OF AN ANDEAN CONDOR

10 YEARS

AVERAGE LIFE SPAN OF A TURKEY VULTURE

GLOSSARY

acid (AHS-id)—a substance that can dissolve some things

bacteria (bak-TEER-ee-uh)—a small living thing

carcass (KAHR-kuhs)—a dead body

migrate (MY-grayt)—to move from one place to another

predator (PRED-uh-tuhr)—an animal that eats other animals

prey (PRAY)—to catch and eat something

roost (ROOST)—to rest or sleep somewhere

toxic (TOK-sik)—poisonous

BOOKS

Nickel, Scott. *World's Grossest Animals.* World's Grossest. Minneapolis: Lerner Publications, 2022.

Riggs, Kate. *Vultures.* Amazing Animals. Mankato, MN: The Creative Company, 2023.

Walker, Alan. *Birds of Prey.* Built to Survive. New York: Crabtree Publishing, 2022.

WEBSITES

California Condor
sdzwildlifeexplorers.org/animals/california-condor

Vulture Birds
www.dkfindout.com/uk/animals-and-nature/birds/vultures/

Vultures Are Gross-ly Important
kids.nationalgeographic.com/nature/article/vultures

INDEX